Sudoku

For Adults

Table of contents

Easy

6		7	4	8	5			1
5	2			9	1	4	7	6
	4	9	6	7		8	3	5
		1		2				4
4	5	3				6		2
9		2		3	4	1		7
	1	4	8	5				3
3	8				7		1	9
2				6			4	8

SUDOKU - 2

Easy

			2	3				1
	1	5	8			7	9	3
7	8			1	9	2	6	4
	2		4		3	9	7	
3	5	6		9			4	2
		4	6		1	3	5	8
4	3	7	9	8		6	1	5
5				7	4		3	9
	9	8			5			7

SUDOKU - 3

Easy

1	9	4	7		3			6
6		3	1	9	8	4		
		8	2	6	4	9	1	
4	5	2	3			6	9	7
	6	9	5	2	7	3	4	1
			6				8	2
			8	3	2	1		9
							5	
2	8		9		5	7	3	

SUDOKU - 4

Easy

		3		7	1		8	9
			2	4	8	6		3
			5		9	2	7	1
	7		8	5			2	4
	6	5	4		7	9		8
		4			3		6	7
4	8	2	1				3	6
5	3				4	1	9	2
	1	7		6		8	4	5

SUDOKU - 5

				4		5	7	8
8	1		2	5				6
7	4	5	6	8	3		1	
2	8		5	7		1		
	6	7	8			2	5	3
9					2	6		
3	2		4	1	5	7	6	9
		6	3	2	9	8	4	
4	9	1	7				2	5

SUDOKU - 6

9	1	3		7		6	8	5
	2	4	5	9			1	7
		5	6			2	4	9
4	6				1	9		
	5		3	2		8	6	
			8	4		5	7	
5	4		9	8	7			3
	3	1	4	6	5	7	9	8
7			1	3		4		6

SUDOKU - 7

Easy

7		4	9		1	5	3	2
			5	7	8			
6	5	1	4					
3		2	6		9			
		9			2	4	6	
4	6			1	5	7	2	9
2				3	4	6	7	1
1	4	7	2	9	6	3	5	8
	3		1				9	4

SUDOKU - 8

Easy

	5		9	8	2	4	1	6
	8	9	3			5		7
		2	7	1	5	9	3	8
			8			2	4	3
2	4				9		6	
	7			5	4	8		1
5	2	4				1	8	9
			5	2				4
	3	7	4		1	6	5	2

SUDOKU - 9

Easy

	1	5		4	9		3	8
8	6				5		4	9
	9	7	3					
	4			6	3		5	
		2	7	9	4	8		
1	7			2			9	4
3	8	1				4	7	2
	2		8	3	7		1	5
7	5	9	4		2	6	8	3

SUDOKU - 10

Easy

3	7				2		5	1
2	1	5	8	6	3			7
8			1	7	5		6	
	6	2			4		8	
		8	5	2			7	4
			9	8	6	1	2	3
5	2	3						
6		9	2	1	8		3	5
	8	1	3	5	9		4	

SUDOKU - 11

Easy

2	8			3	1			
1		7	9	6			5	8
4		5		2	7	9	1	3
9				7	6		4	
7			4	5	8	1	3	9
		3		9			6	7
5		8	7	4	3	6		1
		4	2	1		3	8	5
3	1	2	6				9	4

SUDOKU - 12

Easy

3		2			7	6		1
		1	9	4		3	5	2
	5	4			1			
5	3	6	1	2	4	8		7
8	1	7	3				6	4
			6	7		5	1	3
		8		6	3			
4				8		1	7	6
9	6		7		2	4	3	8

SUDOKU - 13

Easy

	7		3		9	6	5	4
9	4				6	1	2	3
2	6		1	4	5	7		
7	8	2	5	6	3	4	1	9
		6	7		1	2		
5							7	6
	5							
8	2		6	5	7	9		1
	9	1	2	3	8	5		

SUDOKU - 14

Easy

7	9	3		8	5	4		
5	2		4					8
8	4	6	2	1	9	3	7	5
3	5				1	6	2	9
		9	7	2		5		
4				9	3			
			3	6	8		9	4
9	3		1		4	2	5	6
6		4					3	7

SUDOKU - 15

Easy

8	2	4	1					9
		5	8	2	9		7	
	9	7	4			2	3	
			3	9	5		8	
2		9	6		8	1		
5	8	6	7		2		4	3
9	4			3		8	1	6
	5	8	9	6		3		4
	1		2		4	7	9	

SUDOKU - 16

Easy

1	4			5		9		
6	9	8		3	1	5	2	
		5		9		1	4	8
8	7			1	4		3	5
2	1		8	6		4	7	
9	5	4			2		8	
4	3	2		8		7		
7		1		4	6		9	
5	6	9	7				1	

SUDOKU - 17

	3	2		6	7	8	4	
5		9			1	2	7	
4		6	2			5	1	
3	2			7	4			
8		7			2	9	3	5
		1	8	5			2	7
	5		1	2	8	6	9	
2		8		3	6	7	5	
	1	4	7		5		8	

SUDOKU - 18

5					7			
	7	4	5	2	6	8		3
	2	6	1	9	3			
1							5	7
6	4					3	2	9
	5	9	7	6	2	1	8	4
2	6	5	3	7	8	9	4	1
	9		6	4		5	3	2
4	1							8

SUDOKU - 19

		2				8		9
3	9	4		1	8	5	6	7
7	5			4		1		2
	3	1		9	4	2	7	
6	7	5		2	3	4		1
2	4		7	6		3		
	8		1	7		9	2	
5		7		3	9			8
9	1	3	6	8				4

SUDOKU - 20

8					5	3	1	
	2		1	9	4	7	8	5
		5	8		7	9	6	2
	6	1	3				9	
3	9							8
	8			4	1	5		
4			2	5	9	6		3
	3	6	4	1	8	2		9
	5		7	6	3	8	4	1

SUDOKU - 21

Easy

5	1	4	6	9	3		7	
2		8		4	7			3
6			2	5			1	9
9				6	4		2	
	2	6					8	4
8	4	1						5
4	6	5	9	2		8	3	7
3	7	2	4	8	5			6
	8	9	7	3			4	

SUDOKU - 22

Easy

	7		3				5	1
			2		4	9		6
3				1			8	
6	8	1			9	3		
9	4		8		6			2
5	2	7		3		6		8
2	5	9	1	4	3	8	6	7
	6		9		2	5		3
1	3	8	5	6		4		9

SUDOKU - 23

4		8	7	3			6	9
2	6	3	4	1	9	8		
	5	7			8		1	3
		9	3			6	2	
6	3	4	9		1	7	8	
	2	1		7	6		3	
3		2		5		1	4	
		5			2			6
	8	6	1	4	3	5		

SUDOKU - 24

	5	8				1		7
2				6		3	8	4
						9	2	
8	3	2	1			4	7	9
9		1	4	7	8	5	3	2
	4	7		3	2	8	1	6
4		3	6				5	
1		5		8		6	9	
	8		3	9		2	4	1

SUDOKU - 25

Easy

5	7	6			3			4
9	1	4	7		5	8	2	3
							7	
		7	4	2			6	
		5		1	6	4		
6		1	5	3		9		2
7	2	8				3	5	9
4	6	9	3	5		7	1	8
1	5		8	7	9			6

SUDOKU - 26

Easy

			2		3	9	4	5
	5	4	6			2	8	1
9	8			1	5		3	6
	4	9	3			8	6	
		8			2	4	5	3
			8	4	6			7
		1		6	4	3	7	8
	3		7				1	9
	7	6	9	3	1	5		4

SUDOKU - 27

	5	1		2		8	4	3
2	3	4		1	8	7		9
		6	3		4	2	5	1
1	9	8	7	4			3	
5	6	7		3	2		8	4
	2			9			7	6
3		9	4				2	
6			2	8		3	9	7
8						4	1	5

SUDOKU - 28

2			4			5		6
1	6	4	2	5				7
	8	7			1	9		
9	5		8	1				3
4		8	3		7	6		1
7	3	1			2	4		9
8	1	5		4	3		6	2
3		2			6			5
6		9	1	2	5		4	

Easy

6	5	9	8			2	7	
3	2	7	6	5	9	4	1	8
	1			7		6	5	9
7	9	8	3				4	1
5				8		7	3	
	6					9	8	2
	3	1	7	9			6	5
9			1	6		3		4
						1	9	7

Easy

	1		5			8	9	
	9		1		3	2	4	7
	4	8			9		6	
1	6	7	4				5	
8			9				7	6
9	5					4	2	1
	7	6	8	9			3	2
5			2	6	4	7	8	9
2		9	3		7	6	1	4

SUDOKU - 31

Easy

	4		9			1	8	7
9	7		5	8			4	2
1	3	8	2		7		9	
3			4	9	8	5	7	6
	9	7	1	6	3			
8			7			9	3	1
	2	1			9		5	3
6				2	4	7	1	
7		9	3				6	

SUDOKU - 32

Easy

8			2		4	6		5
	2	3	8		5	4	1	7
	5	9			6		2	8
		6	9	5	8			1
						8	6	
3			6	4	2			9
1		7	4			9	5	2
9	6	4					8	3
2	3	5	7	8	9	1	4	

SUDOKU - 33

Easy

	7	3	8	4	1	2	5	6
5					3	1		8
		8	5				7	
7				5	2		6	
8		1	7	6		5	9	3
6	4			9	8	7	2	
2		4		1		9		
1	5	7	4	3				2
3			2	8		4	1	7

SUDOKU - 34

Easy

1		6	8					
	3		5		4	9	6	
			6	2	1	8		4
	6	9		1	7		2	8
	2	7	9	5	8	1	4	6
	8	1	3	6			7	9
8		3		4	5	6	9	2
			1	3			8	
9	5	4		8			1	

SUDOKU - 35

Easy

						6		5
9	7	3	4	6	5		8	
5	6	2	7	1		9	3	
1	2		8			5	6	9
6		9			7	2	4	
8	4		6		9	7	1	3
	5	6	2	4	3	8	9	1
2		4	9		1		5	
3			5			4		7

SUDOKU - 36

Easy

	3	5	7	2			6	9
	2	1				3		
9		6	8			7	2	
6			5	7			1	
	7			4	2	9		6
2	1				9	8	7	5
		2	4	5		6		3
	5	9	2	8	6	1	4	7
	6		9	1		5	8	2

SUDOKU - 37

Easy

2		8		5	9		7	6
7	1		8	6	3			5
6	4	5						
	6			4		5		
9			3	8	6	2	1	
		4		7	5	8	6	
3	7		5	9	4		8	2
4		6	7		2			9
5	9		6		8	7	4	1

SUDOKU - 38

Easy

		5	3				6	
		6	4	7	1		9	
3			6		2	7	8	
6	4	8		1		9		
5	2		8				4	7
7	1	3		9		6	5	8
8	6		9	4	7	5		3
9	5	7	1					6
1	3	4	5		6		7	

SUDOKU - 39

	6	2			8	3		4
8	1		9	3			2	5
7					2	8	9	1
6				5	4	7		
5			2	7	9		8	6
4		7		8			3	9
2		5		6	1	9	4	
1			7	9		2		8
3	8	9	4				6	7

SUDOKU - 40

Easy

7	8	2	6		5	4	3	1
9	6		4	1			2	
	4				2	9	6	5
3			7		1			6
5			9		6	3	4	7
2	7			3	4		8	
6		9	3	5	7		1	4
8		1			9			
4	3		1		8		9	2

SUDOKU - 41

Easy

	6	1	2	3	9	7		
4	7			1	8		5	
9	8	2	5	4	7		6	3
7	9					2	3	1
		6		7	1		8	
	4		3			6		9
	1			8	3		2	4
8	5	7		9	2	3	1	6
	2				6		9	7

SUDOKU - 42

Easy

	9	6	8	7	1	4		
3	1	7		4		9		5
	4	8	3	9	5	6	7	1
4		2				3	5	9
		1	9	5	3		2	
						1		7
7	2			3	8	5		
8	6	5		1		2	4	3
1			5				9	8

Intermediate SUDOKU - 43

5				8		1	7	9
	1	6				5	2	
8	7			5	2			
6	2		7	4		8		3
	8		2		5	4		
1		4		3	6			2
	6			1		9	4	5
		1		6	4	2		7
	5	7		2	8	3	6	1

Intermediate SUDOKU - 44

5	9	3						2
			5	3			4	9
					9	3		5
9	6			7		4	8	3
8	2	4				7	9	1
	1		9		4		5	
4	3	1	6	5				
7	5	2		9	8	6	3	4
6	8	9	4	2	3	5	1	7

5		4		9		1		8
8	6	3	2		1			9
	7	1	8	5	3			
6		8	1					4
1	5	7		8		6	9	
	3	2		6	9		1	7
3		6				4	7	
7		9			5	3		2
2		5		7		9		

		4	8		5	3	6	
8	7		4	6	9			5
5	9	6		1	2			
7	8		6	9		5		
6			7	2			8	
3		9	1		8	4	7	
		7	2		1	8		4
	3				6	1	9	
1		8				6	2	

	4		7	3			2	9
		6		9		8		7
		2			8		6	5
4	6	1	3				9	2
3	2				5		7	
	8	7					3	1
	1	3	6	5	9		8	4
9			8	7				6
6		8	2		1	9	5	3

9	5			6	8	4		
		8	5	2	1	6		9
7	2			4		1	8	5
4								7
1	8	7	3		2		4	
5	6	2	4	1	7	3	9	8
		5		7	4		6	2
2	7		6					
		9		8				

Intermediate SUDOKU - 49

	2		7	4	6		5	
9	4			2			6	
7	3	6						2
		3	6	8	9		1	7
6	1			3		5	8	
8		2					3	6
	7		4	6	8	1	2	
2		1		5		6	9	4
	6	4	1		2			8

Intermediate SUDOKU - 50

	6		8	9				2
8	1	4		2	5		3	6
9			3		6	1	8	
		5		7	8	3	1	
		8	6	9	1			
1				5			6	9
2	4	1				6	9	8
	8	6		1		4	7	3
3		9	8		4	2	5	1

Intermediate SUDOKU - 51

	4	3		2	6		1	
8				3	1	5	6	4
		1	5		4	2	8	
1	6		3	7				
	8	5		6	9			2
2	3	7	4		8			1
9	5		2	1	3		7	6
3	2			8	7		5	
		6	9					8

Intermediate SUDOKU - 52

3	7			8	6	2	1	
		2	3	9	7	8	4	6
6				2	1			9
5			6				9	8
7	3		8					1
9	8	4	2	1	5		3	7
8				6	4		5	2
4	6				2			
	9		7	3		1		

		6		8	1	7	4	5
7	5			2	9			3
3		1	5			2	9	
8	4	5			3		2	
2	1	9		6				
6		7				1		
1	9	8		7		3	6	
5	6		8		2	4		
4	7	2				9		8

2	6		5		1		4	7
5		9						6
	1	4	9	2	6	8	3	5
	7		2		4	3	9	
1	3	5					2	
4			1				5	
	4		8	7	9	5	6	
8	5			1			7	
9	2		4		5	1		

9	7			2			4	
		4	1	7	3	6	2	9
2		6	5		9	7		8
			3	8	1			
7		8	9	5		3	6	1
3				6				5
			2	1		8		7
1	9	7	8		6	2	5	4
8		5	7	9				6

9	5		7	3	2			8
	2	6		8		1		5
	8	4	5		1	3		
				4		5	1	6
	1			5		2		
6	4	5	2	1	9	7		
	7	9	8	2				
			1			8	6	7
1	6				3		5	

3		6		1	2			8
2		9		5	6			
	8		4	3			2	
9	5	8	2				1	3
	3				8			7
4	7		1	9	3	8	6	5
7	9		8		1	3		6
	6		3		9		8	
8			6	7		1		9

9		2		6		1	8	7
4	7			9	8	5	3	6
8		6	1			9	2	
7	2			5				3
3	6		7	2			1	5
	9	5	8			4	7	2
2				8	5	7		
	8	7	6	1		3		
			3	4		2		

		4	7	9		8		6
					4		1	5
			6		5		4	
1	6	3	9	7		4		8
2	5	8	4		3	1		7
		7			8	2	6	3
	9			1	7			4
8		1		4	6	5	7	9
4	7	5	8	3		6	2	

8				3	4	2		9
4				2		5	8	
		9		5			4	3
		1		8	9	3	5	2
5		7	2	1		9		4
9	3	2	4				1	7
1	6			4	2	7		5
3		5	8	7		4	2	
2		4		9		6	3	8

1	6		5		4			9
9		8					7	4
7	4	3	8	1	9	5	6	2
5	7	4	1	8	2	6	9	3
8				5	7	2	4	
	1	9			6			8
	2			9		8		7
3		1			5	9	2	
				2	8	4		5

		2	1	3	7	5		9
	3	1		5		4		
9	7		8	6	4			1
	2			4				5
4	5		9		6	8		2
3	6	8	5	7	2			4
2	9			8	3		5	7
	8	6	7		1	2	4	
7			4	2		6		8

1				8				
9		2	5			8	7	3
4		8		2	6	1	9	5
3	2		6			7		1
8	5		7					9
7		1	2			3	5	6
2	4	7			9	5	3	8
5	1				2	9		4
			4			2		

6				3		8	2	
8	1		2		9			5
9			5		8		4	
			9	1		5	7	2
2	7	5	8	4	3		9	
	9	6		2				8
4			6	5	7		1	
7	3		1		4	9	5	
			3	9	2		8	4

9	7		3				1	2
8		3		6	2			9
6		2			1		7	
1			2	5	7		3	
	2	9	8				5	4
	3	8	9		4	2	6	7
3		7	6		5	4		1
2		1		7			8	6
	8			2	3			5

1	7	8	3				4	5
			8	1	6	7	2	3
		2	4	7	5			
6				3		2		9
2	3		9		4	8	7	1
7		1			8		3	6
		3	6	4		9	8	2
	2	7			9			
9	4	6	2	8	3	5		7

Intermediate SUDOKU - 67

	1			9				5
	6		2	5			8	3
5			7	6		9	4	
2		3		8		1	9	7
	8		1				2	
		5		7	2	6	3	8
	2	4	9		7	3		6
9	7	6	5	4	3		1	2
3	5	1		2	6		7	

Intermediate SUDOKU - 68

5	9	6	8		4		1	
1	7			5	3		4	6
		4						5
9			5	2	8			
	4			1	6		3	
8		2	4	3			9	1
			7	6	2	1	5	4
6		7			9			8
4	2	1	3	8	5		7	9

SUDOKU - 69

4	5		3	8	1	2	9	6
8		6	9		7	5		3
3		9	2	5	6			
1		4			8			
2					3	9	6	
6		3	5				8	1
			1			4	7	5
5	6	2	4		9	1		
	4	1			5			

SUDOKU - 70

			8	4		1	6	9
1	9	8		3	5	7		2
7		6	1		2	5		3
	1	4	2	6	8		9	
6	7	2		5	3			8
3		9					5	6
			7	6			3	5
9	5	7		8	1		2	
4				2				1

			9	5	7	6		
7	6		2	3		9	8	5
	9	2	1	8				
4					3	2	5	7
	2			7				3
	7		4	2			6	8
	4	7		6	2			
		9	3	1		7		
8	3	6			9	5	2	1

		5	7		8		3	9
7	4		5		9	2		
8		2			3	5		1
	8	1	3		5		9	
4		9		6			8	
	3				4	1		5
5	1	8	9	3	6		4	2
9	2		4		7	8	1	
3					1	9	5	6

5			4		2	9	3	
3	4	6			5	2	1	
		9	3	6		7	4	5
		1		5	7	4	9	2
		4			1	6	8	
7	9	8		4		1	5	
	6			7	9	5	2	
9	7	2				3	6	1
4		5		2		8		9

7	1	9	2	8	5			3
8			4	3	9		2	
2			6	1		5	9	
							3	9
5		3	9	7		2	8	
6	9			2	3	4		
		6	1		8		7	
4		8	7	9	6		5	
9	7			5	2			4

3	2	7	5	8		4	6	
9	8	5	3			2		1
6				9		5	8	
7		1	6	5	8		4	
2		6			3	8		5
5			1			6	3	
					6			8
	7	2	4			3	9	
1	6	3						

1		8	4		3		5	2
	2			8	1	7	9	3
5	3		6	2		4	8	
	4			7			1	
8				5	9		6	
9	5		1		2			
	9		7	6			2	8
3			9	1		5	7	
7	6		2	3	8		4	

4	2		8		1		7	6
		3				4	2	
5			2	3				1
3	6	7	5	1			9	
8	9	1	3		7			
	5		6		8	7	1	
6			7	2	5	1		
7	1	2	4	6	9	5	3	8
	4	5	1	8		2	6	

	7	4				8		1
9	5	6	1	8				7
8					7	9		
		3	8	1	9			5
5	1	8	3	7	6	2	4	9
6	9	7	2	4	5		8	3
1	3	5		2		7		4
					4	3		2
		2	7				9	8

Intermediate

6	3	2		4		5		7
4		7				2		1
1	9	5		2		8	4	6
5					7	6	8	3
	2	6		5	8	1		
7	1		6		9			
		4				3		
	5	3	1	9		7	6	
		1	3	8	6	9		4

Intermediate

			8			4		2
7			3	2	4	9	5	6
	2					7	3	8
	9	2	1	5	8	6		4
8	1	7						
4	5	6	2	7	3			9
	6		4	3	1	5	9	
	7	9	5	6				
	4				9			1

	1	4		5	2		6	
	6		8	7		3		
9		3			4	8	2	
3	5	9				2	7	
4	2				7	1		
7	8	1				4	9	6
	4	7		3	8		1	
6	9	8	2	4	1	5	3	
1		2	7			6	8	

8			9		2	4		
		5	8	1	7	3	6	9
1		3		6			8	7
	1	8	2	4		9		
	3	9	1		5		4	2
			6	9				3
	8	7	4			6	9	
		1		8	6	7	2	
2	6	4			9		3	

3	2	9	6	7	8	4	5	1
4	6		1				3	8
				4	3	7		6
						8	7	
9	8	3			7	1	6	
6	7	1		8			2	5
2		5	8	6		9		
		6	9	3	4		8	
	9	4	7		2			3

5	7	1		2			3	
3	2			9	7	5		
9		6	3	5		7	1	2
1		9		4	5	8	2	
4			2	3	1		9	
			9	8	6		4	5
2		7		6			8	1
8		5	4	1	2			6
		3	8				5	4

SUDOKU - 85

1			4	9			3	7
6			7	5		2	8	
5	3	7				4		
			3			6	2	8
				8		1	7	
	7		2		1		9	4
			6				1	3
7				9	8			
	4	3	8		7	9		6

SUDOKU - 86

		4						
2					5	6	9	
9		1	2			7	8	
		2	7	5	4			8
	6	3	1		8		5	
	7	5		3		1		
5	2	6	3	8			7	1
3			5		1	8		6
		8		4				

	6	5			7	4		1
4			3		1		2	
		8	5		9	6		3
	7	1			5			9
	8	3				2		5
		4	8		2		3	
	4		1	5				2
	3		9			1		4
	1			2	3		8	

	8		7			9		3
9		7		2	3			8
	2				9		7	6
	9			1		8	3	
			4	7		5		9
		6	9	3	8	2		7
3	6	5				7		
		4		6	1	3		
	1					6		2

Hard

	6			9			7	5
		9		1	4			3
5						1		
3	1	6		8				
			1		6			
7		8				6	3	1
6					5	9	1	
1		3		2	8	7	5	6
9	8	5	6			3	4	

Hard

5	1	3	6	7				
		4	3		8		1	9
8		9	1	4		7		3
	5		4				9	7
	8	7	2		1	5	6	
			7					1
9			8			1	3	
		8	9	1		4		5
		1	5	3	4	9	2	

SUDOKU - 91

7	5						4	1
	2	1			5	9	8	
					9			2
5	7			1		2		
8	1			6		3	5	
			7	5	3	4		
6	3		4	9	1			5
			5		7	1	6	
		5		8	6			4

SUDOKU - 92

7	2			1		9		
8	9				6	7	1	
	4	6		9	7	5		2
4	8		7		9	6		
			5			4		8
	5			2	8			7
2	7		1	4				6
						1	4	5
		4		8	3		7	9

SUDOKU - 93

4	7			1				2
3			2					9
	2		8		4	7		
					2	6		5
	5		6	9	1	4	7	
8		6	7	3			9	
	6	2					5	7
		4		2		9		6
	1		5	6		3		4

SUDOKU - 94

5		7				9		
			9			3		
9		3	5	1	8	7	2	
8				4		2	6	
			9	3	6		8	
4		6		8	5	1		
2		4	6		1	8	9	
				2	3		7	
	8	1		7	9			

SUDOKU - 95

Hard

3	8		2		1		4	
	7		5	4	6	3		9
						2	1	
			8	3				
					5			3
8	1		7	6			5	2
9	3				8	4	2	1
4	6	8	1	2			9	
5		1		9				

SUDOKU - 96

Hard

	1	6			4	3		2
4	7				3			8
5	2				6		1	
		7						
	4		3	9	5		2	
		5	8		2	9		
7			4	1	9	8		6
3	8		5				9	1
9	6					7		5

SUDOKU - 97

				7	3		2	
4	2	3	5	1	8	7	9	6
		8	9	2	6	5	3	4
3		6		4	7		1	
2							7	5
	9	4	7	3	2			8
5		2		8		3		
		7	1			2	6	

SUDOKU - 98

9	4			7	1			
1				6	5	7	4	
		5	3				9	
	3		4	8	7	9	5	
		7			9	3		
5	8	9	1		3	6		
3		4		9		2		
		2			4	8		9
		1				4		7

SUDOKU - 99

9	1		6	3		5	7	4
4		7		9			3	
		3				2		
1			7					5
7			8	2				3
	2					1		7
2	7				3			9
3		1	9	6	4			
5	4			1	7	3		8

SUDOKU - 100

3		8					6	9
			8		7	4		
	7			5				
					2	8		1
6		9		8	1	5	7	4
1	8	7		9				2
		3	5	4	8	9	1	
	4	1						
7	9	5	2	1	6			

SUDOKU - 101

Hard

	5		4					
8	9			3	6			4
		3	5	1	8	2		7
4			1	5	2	3	7	
1		8			3	5	2	
	3					4		9
	1	7	3			8	6	
			8			9		
9		4					3	5

SUDOKU - 102

Hard

		2			6			4
	6				9		5	1
1		5		3		9		
	3			9		1	6	5
	5		3	6				
	2	9	8		5			7
		8	6		3	4		2
	7	4		8	1	6		3
	1			2		5	7	8

SUDOKU - 103

Hard

9	3		6				7	
6			9		7		4	
	8			3		6		
7	5				3		6	
		6		4	8	5		9
	4		2		6	1		7
	7			2			8	6
	9	8			4			2
	6	5		7		4	1	

SUDOKU - 104

Hard

1	8	9	5		6		4	7
	5		9	3	7	8	2	1
				1			6	9
			1	5		2		
						9	5	
	9			4	3			
9						1	7	3
6		7		9	2		8	
	3	8	4	7				

SUDOKU - 105

Hard

8	5		4	9				
2		4	3		7	8		9
7				8	4	5		
			7	8			9	
	1		3	4	2			
9				2	3			4
4	7	9		2		1		
1	3	8		4		5		7
		6				9		

SUDOKU - 106

Hard

3	5			1	8	2		
		1	4				8	9
8			6	7		3		
9			7	5	3			
	3	7				9		
5		6	2	3		8	7	
4		9			6	1		5
6				9		7		3
1			7		4			

SUDOKU - 107

Hard

8				7	6		5	
1	6	2	3		4		9	
4	5		1		8			3
			5	8	2		7	1
	8	6		1				
2		1	6	4				9
			7	6	9			8
6	9	4			1	7		
7	2		4	3	5			6

SUDOKU - 108

Hard

	6			4		3	1	7
		8		3				
4	1	3						8
		1		6				9
3	7	2		9	5	1		
6			8	1		5	2	
			2	3	8		4	
8	9		6	5		7		2
2		7			4	6		

SUDOKU - 109

			3	9				7
7		2	4					6
9		8			6	2	1	
4	2		8	6				1
5	8		2				6	3
1	6					8		
8	9			7	4			5
2					9			8
3		6	5		8	1	7	

SUDOKU - 110

3	8		2		6			
			5	9			1	3
4	5		1	8		2	7	
9			3		7			
		3	9				8	7
7	4			6		9	3	2
8	1		7	9			6	
			3				2	1
6			5	1	4		9	

SUDOKU - 111

Hard

1	3	5	2				8	
	8	2	1	7	6		3	5
6		7						2
			4		7	1		
			3		5		4	
	4	1		6		5	7	
3	1	4		2			5	6
7	2	8	6		1			
			7	4	3			1

SUDOKU - 112

Hard

4		2	7				3	
6	9	5						
3	1	7			5		9	
	2	8	1	5	4	7		
	5	3	6					1
			3	9	7	5		
	7	4		1				9
	6	1				2	7	
	3		5		2	4		

SUDOKU - 113

5			1					7
3				9	2			
		2					9	
4		7				6		9
9						5	1	
			9	4				2
	9	4	3	2		7		
7	3	1	8			9	2	6
2	5	8	6	7	9	1	4	

SUDOKU - 114

	8			4				
			6		7			
		5		1		8		6
				7	6	1	9	8
9	3	8			4		7	
	1			5			2	
4	5		7	6	8		1	2
7	6	1	5	9		3	8	4
8		9			1			

SUDOKU - 115

	5						1	7
		3	9	5				
1	4		3	6				9
	2	4			9		8	3
	1	9	4	7	8		2	
7	8		2	3			9	4
							6	1
			1	9		4	3	8
5	3					9		

SUDOKU - 116

		8	1	5	3	4		
3	5	4	6					
7		1		8	4	5	9	3
8		7						1
5				2	3			
						6		
	8		4		6	7		
1	7	9	5				2	4
6			7	2	9		5	

Hard

	5				7		9	1
2			1	5				
6		7	3					4
	4	8	9		1	3		5
5	6		8	3				7
9		2	4					8
	9					6		2
1	7				8	4		3
3		4		1	6			

Hard

	3				9			7
	8		3			6		4
6		5		7	4			
					8			6
1	6		5		7		3	
2		8	1	6			9	5
9		6	4		2			1
8	4		7	5				3
3	1		9					2

SUDOKU - 119

9		4		6	1			
		5					7	
6		3	2			5	1	4
5					6	1		
8	2		7	5				
3		6			8	7		2
1		2	9	3	4	8		
			6	8				1
4		8		7		3	9	

SUDOKU - 120

	6	3	9			1		
		2		6	5			
	9	7	8	1			3	
3					6		7	
9		6			1		2	
7		1		8	4			
6			1				4	8
1	3	4		2	8	7	5	9
	5				9	3		

Hard

2			4	3				
6		4		5		9		1
		3		1			4	2
			1	6		5		
7	5	1		9		8		
		6	7	8			2	3
		5	9	4	6		7	
4	9		5					
3		7	8	2	1			

Hard

	1							6
2						4	1	7
3	5		7					
		3	4			9	8	
9	6		1	2				4
	4	1	3	9	7		6	
5			9		4			3
6			8	1	3	5	2	
1		9				6	4	

SUDOKU - 123

Hard

	7	3	9		6			4
					2	3		
	8	6		5		2		1
		7		6		1	4	
4		9		3		6	2	
	6						9	3
	9	1		2	5	4		
3	4					7		2
6	2		4		3			5

SUDOKU - 124

Hard

6			5			4		3
	2							
		4		9	6		8	2
2			3	4	7			
	4		2		1	7	6	8
1			6			3	2	4
	9		7			8		
8	7	2	4		3	6	5	9
		3	9					

SUDOKU - 125

7	5					6		
2			5	4			7	1
6	4			7		2	5	
4	1		2		5	3	9	
		9		1		5	6	8
		5	3				1	
9						1	4	
1			6	3	7		2	
	3	2					8	

SUDOKU - 126

5			4		8	3		
8	3				1			
			2		3		7	8
	2	5			9	1		
					5	4	6	2
	8	3	1	2	6	7		
	5			1		9		7
	1	9				8		6
	4		9	8		2		1

SUDOKU - 127

Very Hard

1	8				6	5		
			5	8				
	4		7		1	9		8
	3		6		8			
8	6					3	5	
						4		
		4				8	7	5
	2	8	9	5	7			3
7		1		3			2	

SUDOKU - 128

Very Hard

					8		2	
5		3	7	4			8	1
8	1				5	4	6	
	8						4	2
3				6				
7	2	6		9				8
		5		8	9		1	
					7		3	
		7	2		1		9	

	4			8	9		2	7
	2	6			5		1	
7	3		6					5
				5	7			9
				1	6	8		3
		1	9			2		
	5			4				2
	9	4		6	2	7		
6	1				8			4

3				1				
		9				5		1
			5		7		4	
	7		2			3		4
5	9		6	7		1	2	
			8	4		7		
				8	9			3
2		7		5			8	
			1			4	6	5

SUDOKU - 131

					7			
			2	5	9			6
2							3	7
9					6		5	1
6		2	5	1	3		4	
	5			4				3
7	1		3		5		2	4
3						1	7	
		5	1	7				

SUDOKU - 132

8	3	7		1			4	6
2				6		3	8	
5		6	7		3			
				5		6		9
		3	1	4		7		8
3					8	4		
7	6	8	4					
		9					7	5

SUDOKU - 133

4	6					5	2	1
			4	6	5	3	8	9
	5	9			1			
7				9			5	
					2		9	
		2	5	8	4			
					7		3	
8				5			6	
5	9						1	7

SUDOKU - 134

	3		7			8	6	5
	1			6				
		6	9		5			
	5						4	
3	7			1	2		8	
		2			7		5	3
	2		6				3	
5			1	8			2	7
1		7	2	5	3	4	9	

SUDOKU - 135

7	4				8		2	1
	8	1	7	2			4	
	6			1		7		
2	9		3		7			
			9	6	4			
						6	9	7
				4		1		
					6	9		5
5	3		2					8

SUDOKU - 136

7						6	8	5
4	5		9		3		2	
			5				3	
3	8							6
			1	9				
	1	4				5		
	4		6	3		1		
			8	2				4
9	7	2	5	1				8

Very Hard

	1		3			2		9
6		7			9	1	5	
				1	2	8		
8	7							
		3						8
	9		7		4		3	2
	4	6	9				1	7
7				6			8	
	8	1						6

Very Hard

	9			5		6	8	
				6				
	6	7		9				4
		3	9		7	2		
	2	9			1	5		3
	8			2				
		6		9		3	7	2
	7			1	3			
	3	2		7				1

SUDOKU - 139

		8	2		9			1
	5	2		7			9	8
	9	1		5	8			
			8				6	
						1	7	5
9			5	6	4	2		
5			9		3	8		
1				8				4
			7					6

SUDOKU - 140

				3			5	
5			2		4	6		
			5			9		3
4		2	3				8	
	6	3			8	7	9	5
8				7				2
			7		3	5		
	4		6		2			
6	5			4				7

SUDOKU - 141

Very Hard

					7	4		
3				6				
		1	9			3		7
8		3		2				
		2				7	3	4
1			4		5		2	8
6	3		2					5
2				4			7	
7		8		1			4	3

SUDOKU - 142

Very Hard

9				1	2			
5		2				4		3
	3				8			1
4			8	3		7		
	5	1			4		9	
8		3		7			4	
3			4			6		
1				6			5	
2	6				7	1		

SUDOKU - 143

Very Hard

		2					4	1
1	6	3	7		4		2	
			9	2		6		3
6			8	4	3	1	9	7
	3		1					
		1	5	6	7	3		2
7	8				5			
			7	6		8	1	9
						2		

SUDOKU - 144

Very Hard

8			1		2			3
9			4		7			2
		5		3				
7				1	8		6	
5		3		6	4			
		1	7	2				
3						4	8	5
	4			7	1	3		
				4		7	1	

	4	6				8	7	9
7	3	1					5	2
	2		7		5	6		3
	7				6			1
2	1			7				
6			5		2			
9		2	3				4	
3								
1	5		9	8	7			

	5				4	2	9	
			9		2		1	
9	4			8	1	3		7
		6	8		7	5		2
	7	3	1		5	4		
								1
2			3	4				6
	6	4	5		8			

SUDOKU - 147

	1	2			7	3		6
			5					
7		6						1
4	8			6		9		7
	5		4	8		1		2
2		9		7	5		3	4
1					3	7		
			7	9		4		5
								3

SUDOKU - 148

	4	7			5			
			6					
	6	5	1			2	7	4
					6	3	9	
6	1		4		3		2	7
		8	2				6	
					2		5	3
	7	3	5		9			
4				6			1	

SUDOKU - 149

6	1	5		8			4	
	8						1	6
4	3			1		7	8	9
7					4	1	3	5
	9	8		5		4		
	4		2		1			
							5	
	6	4		2	5	8	9	
			4			2		3

SUDOKU - 150

Very Hard

		7	9		3	4		
				2			1	9
	3	2				5	6	
4	6		2	3		1		
				5				
5	1		7				8	
			3			8		1
			6	4		7		
7		4	1			6	9	

SUDOKU - 151

		9	2				8	7
	8						6	
6				4				
	2		3	6			7	
3		8			5	6		
	6			8		1	9	3
				5	3		4	
	7		6	2	4	9	5	1
		5	9			7		

SUDOKU - 152

	3	7		5	4	2		
2	1	4	8					6
8	9	5	6		2			1
	2	3		8	1		9	
	8	9	7	6				
4				2		8	3	
3				4	5		1	
							6	4

	4			5	9			
			4		2			1
	3	2					6	
8	7						4	
		4			3		1	
	5		8	4			9	7
3		9	2				5	6
4					1	8	3	9
	1		3				7	4

9		5			6	1		7
1		3		9	4			8
	4		1	2				5
3					9		1	
2				7			5	
		7	3	5	1		8	
	9						6	
			9					3
	3	4	7			8		1

SUDOKU - 155

			9	5		4	1	
		9	2		6			
				8				
6			5		1	9	2	8
	5				3	7		
	9		8					5
	8	7						
9	6		1	3				2
3	1	2	7				5	

SUDOKU - 156

			2	6				
			5	9			3	4
5		4		3		9		
		5	4	1		3		
		6	3		9	4		8
7				8				
			9	2		8		
2			7					5
4		9	1			7		3

SUDOKU - 157

Very Hard

	1	5		2		6		
	6		8			5		3
							8	
	9	4		5	2			
2		6						4
	3					2		1
		7		3		4	1	
6		1		4				8
	8			7	1		6	5

SUDOKU - 158

Very Hard

6			8			7		4
1				2	9	8		
				1				7
3		8						
		6				4		2
			3				6	8
			5				4	3
5	3		1		4	6		
2			7	9		5	8	1

SUDOKU - 159

	7			6	9	4		3
5	9			2				
						1	2	
1		9		4	2		7	5
	4	7		9		6		
3		5			7			
7		2				5		1
4	5			8		2		
	1	6						7

SUDOKU - 160

	6				7	1		8
4	1			9				
		8	3	6			9	
8		6		7				
	5	2		1				
9				2	4	8		
		4	7	5	2	6		
6					9		7	3
1								9

SUDOKU - 161

					4	3		8
	9	4		1	6			
	5		3			1	4	
7			2		9			
	2	6		8		9		
	8	9		7	3		2	
	6	2	5			7		
	7		6	4	1			
	3			9			8	5

SUDOKU - 162

			3					
	2	8	7	5	1			9
7	3		4	8	9			
3	5					7		8
					8			
						6	2	
2	7	3			4	9		5
9	6		5				1	
	8		6	9	3	4		2

Very Hard

				6	9	4		
	4		3			6	2	
7		6		2			9	5
		9					8	
1	2		4	8		9	6	7
8				9	3	1		4
			9	3				
	1			5				
		2	7				1	

Very Hard

	9				5			4
6	1			4	7			
5			9					8
				2		1		
	8	1		3		9		
7			6					3
			4	1				5
		8	5			3	7	
3		5	8	2	7		4	

SUDOKU - 165

4	5			7			9	
		8		1				
9		6				7		3
		3			1	4	5	2
			9		5	3		
7	4		8	2	3		1	6
		7	4				6	
5							7	8
		9						

SUDOKU - 166

			8	2				5
	6		5	1			7	9
				6		1	4	
6		5		2		9		4
			1			7		6
	4			7			2	8
		4			6			7
8		7				9		
			3	7		8		

SUDOKU - 167

Very Hard

6			8			7	4	
1		4	7		6	3	8	2
8	7	3	4					
	5						3	
2				8				
		1	3	6			5	
3		7	2	4		1		
	1			7	8			3
9				1				

SUDOKU - 168

Very Hard

		5		1	4		6	
	9	3				1	7	
	6		7	9			5	
			1		8	4		
5					2			9
	7			5		6	3	8
	8	6	3		1			
	5		9					
4	1							6

Insane

6			4		2	9		7
						8	2	
	8					5		
2				9			8	1
	3			7			6	
		5	8			4		
9				4			3	2
			6	3		1		

Insane

			4	6			5	
						1		
		8	7	2	3			
2								6
			8		5		7	3
7	8			3				
	3	5		4			1	
1		4			2		3	
6								

SUDOKU - 171

5								
				6	1	9		
							5	
	1						7	2
	3							6
			4	2	8			9
	7			3	6			
	9	5		1			2	
1			2		7	8	9	

SUDOKU - 172

	5			7				
		4			3			
		2						
	9				7			
			5	6			2	1
		7			2	5		9
	2	3		9	6			
	1						3	8
	7		4				6	2

SUDOKU - 173

		3						
	4				9		8	5
				6		1		2
		6	3		7			4
		9				8	5	
	1							3
	9	5	8					
	8		9			5		
	3		1	4				

SUDOKU - 174

						8		
2			7	8	9	1		
3	1		6					5
7		6	3		5			8
	4				7			3
	8			9		4		
9							4	
6								
	2			1				

SUDOKU - 175

3		9		5			2	8
			8			3		
			9	6	3			
			1					
9		1			7			
7		2		9				
6							7	9
					1	4		
	7		3		9		5	

SUDOKU - 176

	9	2	6		5		8	
1			9	4				
		4					6	
				2	3			7
7						8	4	
				8			9	
						6		
	1		8	7				9
9			3				5	

SUDOKU - 177

	8		4					9
		9						
			7		1			
		7			6			4
4	6				5		9	
		2		4		7	3	
	3					1		5
				5	9			2
	4		1			9		

SUDOKU - 178

	1		4		7		3	
			5	3			9	1
					1			5
		8			9	7		2
4			3		6	5		8
	6			8				
9	4							
1			7		8			

SUDOKU - 179

			6		7			
8	2	7	1					
			9					
		5			6		9	
	9					5		
3				4		1		2
5		6	3			8	2	
								5
		9			8		1	4

SUDOKU - 180

	3	1			7		5	
2				4	8			
						7		4
3								
		6				1	8	
	8			9	2			
8		2		6	9		3	
			2		4	6		
							7	9

						2		
6		4		9				
				3			7	
4		7			5	8		
9	1							
	6	5	1		9			
7								8
	5	8				9	2	
	2		3		8			4

			3					
	7						3	
	1			8	4		5	
					6			4
8		4		5			7	
						3	8	
9		8					1	7
			2	7	5	9		
6	2	7						

SUDOKU - 183

4		1		3		2		
		6					1	7
				5				
						9		
		8		1				
1	6					4	5	
	1		7		3			6
	5		1	9	8			
9		2						8

SUDOKU - 184

	2	9				3		
6				8				2
	7			9		1		
4			7	5				
	5	6						
						7	1	
						4	9	
9				2	7			3
		5	3		9	8		

SUDOKU - 185

1			8	4	6			
9								
			3	2		5		
				4			7	8
	3			5				2
2	7	4		1				
	8					2		4
6	4						3	1
			6					

SUDOKU - 186

			4	6	2			
	3	7						6
	8		1			5		
		9	3					1
				5	8			
						8	4	
7			8		9			
1	9		5				2	
3		8		2				

SUDOKU - 187

				6			1	7
	3						5	
			4	5				9
				2				
1			3			8	7	
		6	5		8	2		
2					1		3	
		3				5	9	
8			7			1		

SUDOKU - 188

	3					9		4
			9		5			
		5			7			
7						4	2	
	9		1			3		
		3		8		5		7
2					3			
3	5					2		
		9		5		8		1

SUDOKU - 189

	5		1					
8				6	7			
			9				1	4
			8	3	9	7	4	
7	6		5		1		3	
						2		
		9	3		6	4		
	1							
2								5

SUDOKU - 190

		8	9					
		7				9		
		4	2	7	1			
	8	9			2		4	
		1				3		
7							6	
	5		4			6	1	2
			1				3	
4					8			9

Insane

		9	1			4		
				6	5			
		1		3		8	6	5
2			9		3			1
	9	4			6	7		
8								
								7
	8	7				9		
			5	1	2			

Insane

			5	6				2
				4			3	
1				2	7	6		
8				2				7
	5	6	9					
9								
2		9	7					
3		1			5	8		
			1		9	4		

Insane

2			4					
	7				3			4
				1		8		
		5		1				
	3							6
	2	9		7	8	4	1	
	8				9	7	6	
					5			
		3	2		7			1

Insane

		7		2				5
9				8			3	
	2				4		8	
				1				4
				7	5			
7	5					9		
		4	9		7	6		
				4	1	3		9
3				8				

Insane

		3		1				5
				3		4	1	
	5		6	7		2		
			5	6			8	
			8	2			4	
	4	2				1		
4		9						
	6					5	9	3
		7						

SUDOKU - 196

Insane

9			8				4	
			5			2		6
					7		9	
	5				6	1	7	4
		4		2	5			9
		6					8	
						9		
			1					
5	8	7		4			3	

Insane

			7		1			
	6	2			4			
1		7			5	8		
	7				3	9		
9								4
		4				7	8	
	3				7			
8	1					6		7
			2	5		3		

Insane

		7	6					
		9	8	1				
			1		3		5	
	3					9		
4		2		9	6		8	
8						7		1
			4		3	1	2	
				6				3
	2							

SUDOKU - 199

Insane

		4		5				
			3		8			
		5	9				6	
1						6		3
	6			8	5			
		8				9		5
								4
	1				2	5		6
2	4	3			7			9

SUDOKU - 200

Insane

			1				2	
		5					4	
7	6				8			
			6		1			
	4	3		2			5	6
							9	
2			7	6				
	1	8	5					2
		7		1	2	3		

Insane

4	8				3		6	2
3	7				8	4		
		6	5	2				
		7	8		9			
				4			7	
			2		7	3		
						9	5	8
				6				
5		3						

Insane

		2						8
	3						4	
9			4				6	3
6	7			5				
			7					4
5		4				3	9	
			8	9	5			
1		3		4		5		
				1		8		

SUDOKU - 203

	4			8			7	
5	8	3						
			6			8		
					3			
		5	2	6		3		
				4				5
8		6				7	3	
3				2	9		8	
4					6		2	

SUDOKU - 204

2					6			
8				5				4
			8	7		5	3	
	1							
9		3						7
		7		6			1	9
4	9	2		3				6
3			4	8			7	

Insane

	9		2		6		8	
8	5							
		3	4		9	5		
4	7							3
		9			2			5
	3				4	9	7	
	2		7					
6					1			
			6				4	

Insane

8	1		9				7	2
	9					3		
		5	4	3				1
		2						9
7	4	6					8	
							2	
				9			1	
					7	8		
1			2	4			3	

Insane

5	6							9
	8	4	7	1				3
					5		6	
				7	1			
	1		3	4				8
3				2			7	
8							3	
			4					
		2	1	6				7

Insane

9			6				7	
		1		2			4	5
				9	8			6
			2	8		3		
			1				5	9
				5	3			
		4	8					
		2				5	3	8
3								7

Insane

	6					3		
					3	5		1
		1					6	
	2	4	5					9
		8						
1				7	4	2		
		7		3				
					7	9	2	3
	5		1	8		7		

Insane

		7					8	5
							9	
2				3				
	9	2		5				
		5				4		
6	1	8				5	7	9
				7	6		3	
	3					1	6	
		1	3	2				

Solutions and PDF

To gain access to the Ebook version of the test and its associated answer key related to the specified publication, kindly utilize your mobile device to scan the QR code provided below. We regret any potential inconvenience this might cause; however, our assessment revealed that incorporating the answer key directly within the book led to a diminutive font size, posing readability challenges, and contributed to unwarranted bulkiness, thus hampering portability. Opting for the Ebook format empowers you to conveniently magnify the content and significantly lessen the book's overall weight, resulting in a more user-friendly encounter. Your comprehension and collaboration regarding this matter are deeply valued and acknowledged.

Easy   Medium   Hard

Very Hard   Insane

Made in the USA
Las Vegas, NV
25 September 2024